Bioelectric Vitality

Bioelectric Vitality

Exploring the Science of Human Energy

Richard H. Lee, P.E.

China Healthways Institute
San Clemente, CA

China Healthways Institute
115 N. El Camino Real
San Clemente, CA 92672, USA
(714) 361-3976

Library of Congress Catalog Card Number: 96-92923
ISBN 1-889983-04-7
Printed in USA

NOTICE: This book does not intend to diagnose disease nor to provide
specific medical advice. Its intention is solely to inform and to educate.

To my son, Tomu.

Table of Contents

INTRODUCTION

Human vitality is quantifiable. For thousands of years, Chinese doctors have insisted this was so, but western scientists have investigated these claims under the microscope and repeatedly found nothing. Even with X-ray, MRI, and a host of high tech equipment, most scientists continue to insist that human vitality is nothing more than the product of a vivid imagination.

For several decades, Kirlian photography has been used to photograph living things, usually leaves or fingertips, and the results have pointed toward some aspect of human energy affecting the images, but an acceptable framework for understanding the phenomenon has not yet been presented.

Actually, that framework has existed for thousands of years as the basis of traditional Chinese medicine. It took me a decade of observing the daily application of these principles in my wife's acupuncture clinic to realize that the high curative effectiveness of acupuncture and Chinese herbal therapy are strong evidence that the theory behind them might have close ties to reality. Finally I started considering the principles of traditional Chinese medicine as literal facts, rather than convenient analogies, as most Westerners view them.

Yin, Yang, Qi, and Shen: All are described as vital substances from which our physical world is constructed. As I studied these substances, it gradually dawned on me that I was observing them in the thousands of Kirlian photographs I had taken. And in my photographs there is evidence that they behave just as they had been described for centuries in Chinese literature!

Contemporary western science and ancient Chinese science do indeed describe the same world. Kirlian photography provides a bridge between them so that the West can penetrate further into the secrets of human health, and perhaps even some secrets of human consciousness which have remained hidden from the West within the theory of traditional Chinese medicine.

- Richard H. Lee

CHAPTER 1

THE CHINESE SCIENCE
OF VITALITY

Qigong is the Chinese science of vitality which covers a vast territory. It is an ancient system of integrating the human body with the universe. The practice of Qigong involves abilities which go well beyond what we have previously defined as "normal" and "possible." It also offers a window through which Westerners can begin to understand the phenomenon of vitality. Specifically, it provides a framework within which we can understand the electromagnetic phenomena which produce the images seen in Kirlian photographs.

Qigong Masters work with Qi. Unfortunately, this Qi is that aspect of the Chinese culture which makes the Chinese "inscrutable" to outsiders. Literally Qi translates as air. Qi is often translated as bioenergy and electromagnetic energy. It is also translated as every imaginable kind of energy. The Chinese consider oxygen and steam to be forms of Qi because they contain significant amounts of energy.

Why is it that oranges can sit in a bag for weeks with every orange looking almost as fresh as when they were first picked? Then an ugly green mold attacks one of the oranges. Within a day or two, that orange is covered with the ugly green stuff. Logically, it would seem that the oranges in direct contact with the moldy orange would immediately be infected as well. However, as you examine the oranges closest to the bad one you find that they're OK, but an orange or two on the other side of the bag have started to get moldy. From the standpoint of Qigong, each orange has Qi, an organizing field or coherence, that maintains its defenses against invaders like mold. Those with the strongest vitality remain highly resistant to the mold's attack for months after being picked. Those with weak vitality are defenseless against its attack.

Qi and Western Science

Qi makes more and more sense to western scientists as theories like quantum physics and relativity enhance our understanding of the universe. Since Einstein introduced the theory of relativity in 1905, modern physics has adopted the view that mass and energy are both made up of the same stuff, which can take the form of either mass or energy. Einstein's famous equation $E=MC^2$ shows that mass and energy are interchangeable. The truth of this equation was illustrated in the explosion of the atomic bomb that brought an end to World War II. Qigong Masters agree with this idea. In fact, many say that not just mass and energy, but *everything* is Qi.

Three hundred years ago, Sir Isaac Newton established the fact that a body in motion tends to stay in motion. We believe this even though any object we put in motion tends to slow down and stop. Friction, a vague term, took care of the huge differences between Newton's laws and the real world. Friction causes real world things to slow down and stop. Ignoring friction, Newtonian science saw the world as a clock ticking away time at a steady rate, never stopping.

Thermodynamics, a science that came out of the invention of the steam engine, clarified the nature of friction. Yes, things tend to slow down, to come to equilibrium, but with thermodynamics we discovered that we could extract work from this slowdown.

Thus, the industrial revolution was born. Along with this opportunity to extract energy came the depressing notion that everything (including people as we age) is doomed to degenerate toward dust.

This has always hit me as odd, because, if you look around, you don't see things tending toward scattered dust. Dead weeds eventually disintegrate to dust, but they are promptly replaced by new plants and trees with their ever-increasing organization. And in outer space, while we can find occasional asteroid belts which might be viewed as a sign of accumulating dust, we see brightly lit suns and well-organized solar systems like our own. The universe seems to have moved from the chaos of the big bang toward organization, despite the predictions of thermodynamics.

Back on earth, we see the formation of highly structured crystals, and acorns acting as the seeds for tremendous organization into trees. We see ants working away to create homes to live in. And, unless you live in the wilderness, your world is dominated by man-made organization: cars, houses, neighborhoods, highways, school systems, state governments and international peacekeeping. Man's insatiable urge to organize appears to conform to a universal principle which governs the structures of crystals, plants, animals, and the heavens.

Basic Components of the Universe	The Three Human Treasures	The Three Universal Treasures
Yang - The Electric	Shen -The Qi of Spirit	Heaven - Spirit
Life - The Meeting Point	Qi - Vitalizing Energy	Human - The Vitalizer
Yin - The Magnetic	Jing - Prenatal Essence	Earth - Unvitalized Matter

The branch of Qigong most accessible to Westerners is medical Qigong which involves all forms of Qi that bring life to earth. By understanding the Qi in your body you can get a sense of how Qi works in the universe. In medical Qigong, Qi is generally limited to "human Qi", which includes electromagnetic energy, chemical energy, sound energy, and heat energy. Although they are not easily measured by western science, they are clearly defined in traditional Chinese medicine:

Qi of the Body	Forms of Vitalizing Qi
Yuan Qi	Qi of the kidneys for reproduction
Zhong Qi	Qi of the lungs and heart [food (Gu) plus air (Da)]
Ying Qi	Qi of the blood which nurtures internal organs
Wei Qi	Protective, defensive Qi of skin and blood vessels
Wai Qi	External, emitted Qi, projected beyond the the skin

The Qi we get from food and air is critical to building the above forms of Qi within our bodies. This is why Qigong doctors are so insistent on the value of freshly grown and harvested food, which still contains much of the vitality it manufactured before harvest, and the value of fresh air, vitalized by the sun. Processed foods and stagnant air contain little Qi.

Traditional Chinese doctors believe they can evaluate these energies through such traditional techniques pulse and tongue diagnosis. Then they adjust and balance them within the human body through such methods as acupuncture, herbs, and Gua Sha. If these forms of Qi are real, their effects should be measurable scientifically.

Much of the western study into the function and nature of Qi has focused on the meridians, the lines of conductivity along which acupuncturists direct Qi with their needles. These meridians are measurable electrically conductive pathways within the body. Acupuncture points, where needles are inserted, are points of high electrical conductivity on the surface of the skin which connect to these meridians. From the standpoint of physiology, these meridians do not exist. There are no little wires to be found when the body is cut open, but high electrical conductivity is measurable.

You don't have to be a healer or a doctor to study medical Qigong. There are many popular forms of practice, and all allow the practitioner to study the actions and principles of Qi:

Forms of Qigong Practice

Kung Fu (Gongfu) is Qigong as a powerful martial art. Practitioners can build health and longevity while increasing agility, self-defense skills, and deadly punches.

Tai Chi Chuan (Taiji Quan) is the slow-moving martial arts exercise performed in the early morning by millions of people in parks throughout China, a very popular medical Qigong. These movement exercises, in addition to promoting circulation, strength, and cleansing in the physical body, teach practitioners to feel the Qi that surrounds them as they move through it.

Breathing exercises designed to collect Qi can take a wide variety of forms from walking to sitting or even lying, and build vitality at all levels.

Relaxation exercises reduce tension allowing Qi to flow more freely. When Qi flows all the organs in the body become stronger and more productive.

Quiescent Qigong involves sitting quietly, which allows practitioners to escape from the constant stimulation of daily life and experience the flow of Qi within themselves. This is an advanced Qigong, for it involves withdrawing from the body and working directly with consciousness.

In fact, Qigong is defined so broadly that, if we overlook the tendency to limit Qigong to "things invented in China" we find that Qigong can include a wide range of disciplines including biofeedback, yoga, relaxation, contemplation, and meditation.

Qigong School	Guiding Focus	Exercises
Taoist Qigong	Follow the Tao, inner spirit	Cultivate the inner essence
Buddhist Qigong	Let go of attachments	Quiescence, inner calm
Confucian Qigong	Serve family & community	Build vitality to better serve
Medical Qigong	Heal the sick	Skills with emitted Qi

Oddly enough, if I go out in a park in China and start doing a Qigong form, the Chinese will stop and stare. They might whisper among themselves, "Look, Ma Sha, a Big Nose" -- (their name for Westerners when nobody's listening) is trying to do Qigong. The Chinese view Qi as the interface between body and mind. Thus any exploration of Qi is incomplete without considering consciousness.

Shen as Consciousness

Shen, the Chinese word for spirit, mind, or consciousness is a highly causative form of Qi. Other forms of Qi act to bring the will of consciousness into physical existence. This is not a new theme to the West. "Mind over matter" and "the power of positive thinking" are established western ideas. The uniqueness of the Qigong perspective is that Qi is viewed as an interface through which information is exchanged and through which the mind creates in the physical world.

Qigong Masters will often argue that Qi is the medium through which consciousness interacts with the world. Sights, sounds, and other sensations are transferred to consciousness through Qi. And consciousness acts on the physical world through Qi. A Master spends decades of his life establishing coherence with Qi, strengthening this interface.

Scientists spend their lives studying the nature of the physical world. While they find themselves surrounded by the products of consciousness, they never see the consciousness itself. Even the most advanced scientific theories, while explaining the world we live in with great detail, do not describe consciousness.

Superstring theory offers to bring together relativity, electromagnetism, and quantum mechanics into one consistent formula, that of strings which vibrate in ten dimensions. Einstein's four-dimensional space is hard enough to visualize, so please don't try too hard to visualize ten-dimensional space or your brain may shift to an unrecognizable new configuration! Superstring theory appears capable of predicting all known experimental outcomes.

Like all previous scientific theories, it predicts results, but falls short in explaining the basic causes of these results.

If consciousness is a part of the ten-dimensional universe described by superstring equations, then we will discover that it is one of the variables in the equation. If not, then consciousness must live outside of the universe described by physics. Perhaps consciousness is the sound that plays upon the superstrings. Perhaps the universe is the instrument and we are all members of the orchestra. And Qigong may be our way of becoming better musicians.

Key Points from Chapter 1:

1. Qigong is a broad collection of Chinese technologies for improving health, increasing human potential, and more strongly influencing the world around us.

2. While Western medicine and medical sciences do not yet embrace these methods, breakthroughs in understanding, such as those presented here with Kirlian photography, are rapidly opening the door.

3. Those who wish to explore are finding Qigong to be a tremendous Chinese contribution to western medical science.

CHAPTER 2

MEASURING QI
WITH KIRLIAN PHOTOGRAPHY

Before I began measuring Qi myself, I had read of scientific experiments on Qigong in which a test subject's hands produced north and south magnetic polarities, and others where "micro-magnetic particles" were emitted from the palm of a hand. One experiment showed a Kirlian photograph of the hand of a Qigong Master before he emitted Qi. It was surrounded by a corona of electrical discharge, as if he were hooked up to 30,000 volts and it was shooting out of his hand. The second Kirlian photo showed the Master's hand during Qi emission. The discharge was twice as big and much brighter. These photographs launched my research into Kirlian photography to demonstrate the quantifiable changes in a person's Qi.

Enlargement of a Kirlian photo shows a typical Kirlian image surrounding a finger tip indicating strong balanced Qi.

What is Kirlian Photography?

Kirlian photography was discovered in 1939 by S. D. Kirlian, a Soviet engineer. It is a photographic technique in which a high voltage (30,000 or so volts) is applied to a metal plate behind a photographic film. People place their fingers or other objects on the film and the high voltage is applied to the fingers for a fraction of a second. The result is a corona discharge, a circle of light surrounding the finger which is produced as electrons jump back and forth between the film and the finger, ionizing the air. The light recorded on the film becomes the Kirlian image. Here's how a Kirlian camera works:

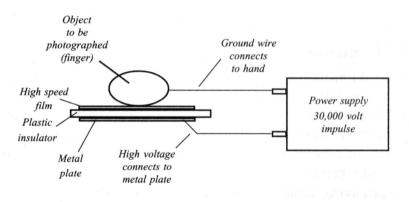

As electrons jump from the film *to the finger*, they group into streams that flow to the finger, making streamers. And, when the electricity flows the other direction, electrons come *off the finger* and form into pools rather than streamers, making balls of light. The amount of electricity that flows between the hand and the film is determined by the electrical conductivity of the arm and hand. Higher conductivity allows more electricity to flow. I found that about five times as much electricity flows in a person who is healthy and strong, as in a person who is weak. The clear conclusion is that increased Qi in the arms and hands allows more electricity to flow and thus produces brighter pictures. Qi makes the arm and hand more electrically conductive. A Qigong Master can move Qi from one part of his body to another at will. By directing strong Qi into his hands, a Qigong Master can make his Kirlian image much brighter.

Many studies have been done which show that the Kirlian image of a leaf, which would normally gradually fade after being picked, brightens dramatically when a healer or someone with a "green thumb" (for whom plants grow wonderfully) simply holds his hand about six inches above the leaf for a few seconds. This is one of the most commonly repeated experiments in Kirlian photography, and supports the hypothesis that healing energy can be transmitted over a distance *without* physical contact. When people who are hopeless with plants hold their hands near, they often cause the Kirlian image of leaves to fade away.

A Controversial History

From the beginning, Kirlian photography was surrounded by controversy. The photographs showed spectacular images of living things surrounded with beautiful patterns and colors. They made it look as if an energy field surrounded all living things. This was, of course, a boon to vitalists like Obiwan Kenobi who insisted that an all-pervasive energy such as Qi was what gives life to all living things. It was, however, quite disturbing to western scientists who believe that life can be explained adequately without such "vague, poorly defined, and unverifiable" theories as Qi.

Scientists who spoke of vital energy, of course, were effectively run out of town. They received no funding for research. This left two other groups. One group did not talk about vital energy, but was open to the possibility of its existence. Some in this group received research funding for many years. The other group felt strongly that the concept of vital energy was invalid. These skeptical researchers successfully convinced the scientific community that results of Kirlian studies were so full of apparently random variables that no research was possible. They vehemently rejected Kirlian photography.

Moisture on the Skin

A variable that many skeptical researchers zeroed in on was perspiration. It is an obvious variable to use in evaluating Kirlian photography because moisture on the hands is an excellent electrical conductor. Varying this conductivity can, in some cases, create substantial changes in brightness of the Kirlian image. Skeptical researchers concluded that variations in moisture of the fingers explained the phenomenon of Kirlian photography, claiming it was the "cause" of the variability.

This is a classic case of researchers ignoring the truth and, instead, finding what they are looking for. These researchers had test subjects do aerobic exercises until they were covered with sweat and found that their Kirlian images consistently got much brighter. Their conclusion: *Moisture makes the Kirlian image brighter.* Other researchers saw the consistent disappearance of Kirlian images of people's hands placed in plastic bags, or dipped in water. The conclusion from these tests: *Moisture makes the Kirlian image disappear.* Both groups of researchers concluded that moisture on the skin was such an overwhelming factor affecting the Kirlian image that Kirlian photography could not possibly be a useful measure.

You would think, when these researchers got together, that they would argue about whether the moisture caused the Kirlian image to get *bigger* or *smaller*. However, they set aside their differences and agreed with each other that their initial hypothesis was correct. "Moisture is such a big factor as to make Kirlian photography hopelessly error-prone." They found the answer they were looking for: Moisture had explained away the phenomenon of Kirlian photography. Important research on Kirlian photography was abandoned simply because scientists didn't see that moisture, by itself, had very little to do with this phenomenon.

In 1976 the death blow for Kirlian photography was issued in an article in SCIENCE entitled *Image Modulation in Corona Discharge Photography: Moisture is a principal determinant of the form and color of Kirlian photographs of human subjects.* It concludes: "In general, the photographic response to moisture suggests that corona discharge photography may be useful in the detection and quantification of moisture in animate and inanimate specimens through the orderly modulation of the image due to various levels of moisture." In other words, the only virtue of Kirlian photography is evaluation of the effects of moisture.

Qi Explains Kirlian Findings

If the existence of Qi is accepted, then the apparently contradictory findings in the moisture experiments fall into place. Increased Qi increases electrical conductivity of the body. This increases the brightness of the Kirlian image. Increased Qi also provides better control over sweat glands. This usually means drier hands. While perspiration has little effect on the brightness of the Kirlian image, the correlation between dry hands and bright Kirlian images is due to high Qi. A person with low Qi doesn't have enough Qi to control sweating, and therefore, often has cold, clammy hands. The principal cause of his small Kirlian images is low Qi, not high moisture. This is clearly evident because test subjects who perspire after aerobic exercise have bright Kirlian pictures. Athletic breathing collects considerable Qi. And profuse perspiration in such cases is due to normal metabolic response rather than low Qi.

Putting the hand in a plastic bag cuts off the oxygen supply to the skin and increases CO_2. This decreases the Qi available to the hand. As expected, the Kirlian image decreases in brightness.

In one experiment, I had test subjects wash their hands in regular tap water, then wash their hands in the same water, except with magnets inserted into the water flow. Each time, for consistency, they dried their hands very well with the same towel. The following Kirlian photographs are typical of what I found: While the tap water decreases the brightness of the Kirlian image, "magnetized" water increased the brightness. Thus, even the nature of the water affects the experiment.

Before washing   *After mag wash*   *After plain wash*   *After 2nd mag wash*

*Kirlian photographs of hands washed in tap water
and magnetized tap water.*

Qi and Spontaneous Perspiration

Cold, wet hands indicate low Qi. Since early childhood I remember my embarrassment on shaking hands with my cold clammy palms. One of the first things I noticed in my practice of Qigong was that my hands became much drier. Qigong Masters consider dryness of the hands a measure of Qigong training. Thus a handshake determines whether a student has been diligent with his practice. Today, when I neglect Qigong practice, I am again embarrassed by my moist handshake and reminded of my neglect.

Searching for Kirlian Consistency

I purchased my first Kirlian camera in 1988 and promptly discovered why researchers became frustrated with Kirlian photography. There were numerous controls: dials to adjust voltage and frequency, and to set pulse duration. And there was a switch, which, if thrown, would change polarity and alter the nature of the pictures entirely. Rewiring the unit allowed me to keep the dials permanently set in one position.

This camera was designed to be used in a darkroom with photographic paper and developing chemicals. Test subjects had to go into the room to be photographed. I viewed this as most impractical because most people don't like dark rooms and I didn't like the idea of dealing with chemicals to develop pictures one at a time. These problems were solved by building a black bag around a Polaroid film holder and attaching it to the top of the Kirlian camera. This produced a unit that could be used in a clinic, trade show, or anywhere.

As I started to use the modified camera I quickly discovered that the electrical discharge varied from picture to picture, giving me variations in readings because of the quirkiness of the electronic circuit. I was tempted to follow the lead of researchers who threw out Kirlian photography because of its unpredictable variability. As I dug into the circuit, however, I found many causes of variability, most of which arose from trying to control 30,000 volts in a small box. It became evident that electrons were carrying such a high charge that they, like the proverbial elephant, could go almost anywhere they pleased. My job was to direct them into the same place every time.

I changed the circuit to run on batteries, and provided a ground electrode to be held by the test subject. This reduced the effects of proximity of the test subject to metal and other conducting media, and reduced the chance of high voltage zaps from accidentally touching a grounded appliance during photography. I also used very high speed film and dropped the exposure time to less than a millisecond so that the test subject would hardly feel the discharge. The resulting Kirlian camera was light, portable, safe, and consistent from picture to picture. At last, I was ready to start taking Kirlian pictures of others.

The Safety of 30,000 Volts

How can a device that runs 30,000 volts through the body be safe? I had seen a fibrillator that was built for surgeons to test implanted defibrillators. It produces a weak (3-10 volt) 60 Hertz signal. When applied to the heart, it induces 60 Hertz activity in the heart which overwhelms the normal cardiac rhythm, throwing the heart into fibrillation.

If three volts can throw the heart into fibrillation, how can 30,000 volts be safe? One important factor is current. The current of most Kirlian cameras is very low, in the microampere range. My camera, with high speed film, uses even lower current. The second more important factor is that the fibrillator produces a repeating frequency, 60 Hertz, a discrete frequency which competes with the natural frequency of the cardiac rhythm and entrains the heart to run at an incompatible frequency. A Qigong Master might say it this way: When the cardiac rhythm is forced into entrainment with an external 60 Hertz signal, the coherence of the heart is disrupted.

Unlike the fibrillator, my Kirlian camera produces a single pulse containing a broad spectrum of frequencies. There is no single frequency to entrain the heart. In all my research and in the volumes of studies I have read on Kirlian photography, safety has never been an issue.

When my Kirlian camera was finally producing consistent pictures, I began taking photos of the most convenient test subject available, myself. To my surprise, I discovered that my Kirlian image never seemed to change. I put my hand in the freezer to make it really cold. I put my hand in water. I tried exercise. No matter what I did, I couldn't see much change. I was having the opposite problem from the researchers who had rejected Kirlian photography. I was ready to throw away my Kirlian camera because the pictures were all the same!

Burned Out Massage Therapists

I began to use the Kirlian camera on Dr. Yuan Zhi Fu, and her patients, many of whom were also professionals in the healing arts. I noticed they shared similar energy patterns. They suffered from aches and pains and infections and other ailments, but especially low energy. I wondered whether this might be related to the work they were doing. Some therapists became so exhausted that they could provide massages to only one or two patients per day. Could it be that, simply by touching a patient, they were transferring Qi? And the question that excited me most of all: Could I measure this loss of energy through Kirlian photography?

I found that, though these massage therapists usually had very faint Kirlian images, sometimes their images were brighter.

Through questioning I discovered that the key variable was how recently they had given a massage. Those who had just done so almost invariably had faint images, while those who had not given a massage for a day or two often had brighter images. I regard this as an important medical discovery because it relates to all practitioners of the healing arts who touch patients with the intent to comfort. They become drained, a phenomenon called nurses' burnout.

Transferring Qi Through Shoulder Massage

I experimented with this phenomenon of Qi transference and developed a surefire experiment which I use at our Life Enhancement Workshops. I choose eight individuals from the audience of doctors and massage therapists and start by taking their Kirlian photographs. Four sit in chairs with the other four standing behind them, massaging their shoulders and trying to soften any knots of tension. After the brief massage, I take their Kirlian pictures again. About 90% of the participants who receive a massage get much brighter images and about 75% of the participants who give a massage get much smaller images.

| *Patient before massage* | *Patient after massage* | *Therapist before massage* | *Therapist after massage* |

The results are highly consistent, making for a persuasive demonstration. Usually, when a massage recipient fails to get larger images, the massage giver fails to get a smaller image as well. This makes the demonstration even stronger, because it appears that, while Qi is usually transferred during massage, it is not always transferred:

I found that those who maintained brighter images even while making another's image brighter often practiced a breathing technique to replenish their energy while giving the massage. Only a few of those who claimed to have a method of "protecting their energy" succeeded in keeping a bright Kirlian image, and even fewer succeeded in making the recipient's image brighter as well.

The effect of energetic breathing toward sustaining a therapist's energy can be studied scientifically through Kirlian photography, and different energizing effects can be tested. Thus, Kirlian photography as a teaching tool can guide practitioners as to whether they are practicing the skill effectively. With the predominance of nurses' burnout, such a study holds immense value. If nurses and other hands-on health practitioners can learn to rebuild their own energy, this will be a tremendous boon to the health of practitioners. Where they can enhance a patient's energy while maintaining their own, either through meditation techniques or by using energy enhancement techniques like Chinese herbs or the Qigong Machine, there will be a revolution in health care. *The key is quantification of Qi, and Kirlian photography provides this.*

Exchanging Massages

If you have access to a Kirlian camera I suggest you try the following experiment. It illustrates several concepts:

-- The draining effect of massage

-- The importance of collecting Qi during treatment

-- The concept of adding Qi to a relationship

Find a partner and take both his and your pictures. Then give your partner a gentle massage on the shoulders, finding and relieving knots of tension. Stop after five minutes and take both of your Kirlian pictures again. You will probably find that your Kirlian image got smaller and your partner's got larger. Next, trade positions and let your partner massage you for five minutes. This time, when you take Kirlian pictures you will probably find that your partner has returned the Qi to you and then some. While her Kirlian picture is probably considerably smaller, you will probably find that overall brightness of the two pictures is a little higher than at the start.

Now, trade positions two or three more times, each time giving or receiving a massage for five minutes. You will probably find that, while the Qi gets passed back and forth, the overall brightness of the two pictures gets stronger. To be fair, be sure to stop at a point where both people have gained equally in brightness.

Enhancing Qi Through Qigong Exercises

A second experiment you may wish to try involves using Qigong breathing techniques. Again find a partner and take pictures, then start massaging his shoulders. During inhalation, visualize energy flowing from the sun and stars, through the top of your head and into your lower abdomen. This collects Qi. Then feel the energy flowing from the lower abdomen, up through the heart, and out the hands during exhalation. If you succeed in collecting Qi, your Kirlian image will get brighter, and if you succeed in transferring Qi, your partner's image will get brighter. Hopefully, both images will get brighter.

Once you have done this exercise successfully, you will be a different person, for you will know that you are capable of collecting Qi and giving it to others. You may realize that you have been depending on the Qi of others and no longer need to. You have tapped into the universe, an unlimited source of Qi.

Adding Qi to any system, whether it is a person, a family, a company, or a world, makes things work better and better. Masters recommend that just one person in a family practicing daily Qigong will strengthen the entire family in health, mood, money, and harmony. This would be an interesting hypothesis to test through Kirlian photography. Companies, communities and nations may also respond in similar ways. With 100 million Chinese now practicing Qigong, collecting Qi daily, it should not be a surprise that China is gaining strength so quickly.

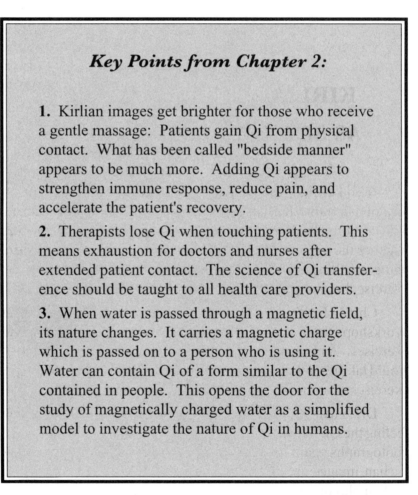

Key Points from Chapter 2:

1. Kirlian images get brighter for those who receive a gentle massage: Patients gain Qi from physical contact. What has been called "bedside manner" appears to be much more. Adding Qi appears to strengthen immune response, reduce pain, and accelerate the patient's recovery.

2. Therapists lose Qi when touching patients. This means exhaustion for doctors and nurses after extended patient contact. The science of Qi transference should be taught to all health care providers.

3. When water is passed through a magnetic field, its nature changes. It carries a magnetic charge which is passed on to a person who is using it. Water can contain Qi of a form similar to the Qi contained in people. This opens the door for the study of magnetically charged water as a simplified model to investigate the nature of Qi in humans.

CHAPTER **3**

KIRLIAN PHOTOGRAPHY
AS SCIENTIFIC EVIDENCE

If you decide to take up Qigong, how will you know if you are doing it right? Kirlian photography provides Qigong students with a biofeedback tool so they can determine for themselves whether the exercise their teacher has assigned to them is really doing anything. They will probably discover that it's not the exercise, but the way they do it that determines success.

China Healthways Institute has offered Life Enhancement Workshops for years, during which we have taught Qigong exercises. As I pursued my research into Kirlian photography, I would take photographs of participants before they began Qigong exercises in a pine tree grove (known for strong available Qi).

Dr. Yuan Zhi Fu then led participants in collecting and feeling the Qi. When their exercises were complete, I took Kirlian photographs again to show participants how much brighter their Kirlian images are after they have collected Qi. Interestingly, Kirlian photography provided persuasive evidence that the electrical conductivity had dramatically, regardless of whether they experienced any feeling of Qi.

Without Kirlian photography, there would be disagreement among participants about what they had experienced. Some would insist, "Wow! I really felt electricity in my hands" or "I feel full of energy" or "I could feel my whole body spinning." Others would disagree, saying, "I didn't feel anything. You guys are crazy." I would tend to agree that this experiential stuff is pretty vague. That's why I like Kirlian photography. Through the Kirlian photos, everyone can see that the Qigong exercises work. Qigong does cause a significant increase in the body's electrical conductivity.

Kirlian and Cancer

In my search for a scientific understanding of Kirlian photography I came across two books on the topic by physicians who were exploring its relationship to cancer. Unfortunately, both used proprietary Kirlian equipment and did not disclose the secrets of their methods. I did deduce some interesting parallels in the ways they used the Kirlian photographs to diagnose cancer. Both focused on the streamers of the Kirlian image.

This Kirlian photograph shows many long streamers, which some researchers have found are indicative of cancer. In my research these long streamers indicate electron (Yang) deficiency.

In one method fingers were wiped with alcohol, then photographed. For healthy people, the streamers—the bright spikes which are caused by electrons streaming in toward the fingers—would disappear for a while. For people with cancer, the streamers would return immediately.

In the other Kirlian method, a pattern of balls was found surrounding the cancer site (indicating an abundance of electrons around the cancer) whereas the rest of the body showed a pattern of streamers, shooting straight out from the fingers.

My Kirlian photographs of cancer patients showed very weak images so it was hard to determine whether I was observing more balls or more streamers. Both balls and streamers are prominent landmarks in most healthy individuals but were absent in the cancer patients I had photographed. I felt that this lack of brightness indicated simply an absence of Qi. These two cancer researchers might have used higher voltage than I had to get a brighter picture. Or it could be that they were diagnosing cases of cancer early, before Qi was severely depleted.

Since long streamers result from a flood of electrons flowing from the film to the finger, the research of these two physicians suggested to me that cancer patients must be deficient in electrons. Had the high-voltage pulse provided an opportunity for the body to absorb the electrons that were deficient? But then why would cancer patients be deficient in electrons? I consulted with doctors who explained that cancer patients tended to have an acid pH which is chemically similar to a deficiency in electrons.

However, Kirlian photographs of the cancer site itself did not show streamers like the rest of the body. The cancer, when photographed, showed mostly balls. It appears that the cancer carries a strong negative charge, an abundance of electrons. Could this mean that the cancer drains electrons from the rest of the body? In terms of Chinese medicine, Yin is often described as the female, magnetic, passive aspect, and Yang the male, electric, active aspect. Thus, a deficiency in Yang (electrons) suggests a low activity, and an excess in Yang suggests an excess in activity.

This is exactly what we see in a cancer patient. Lack of vitality is certainly a classic symptom of the cancer patient, yet rapid, often explosive growth often characterizes the cancer itself. Perhaps this researcher's photographs reveal a mechanism by which cancer grows so quickly while the rest of the body withers away. This may be an important clue to the nature of cancer.

Guangxi Tumor Hospital

I accompanied Dr. Fu to Guangxi Medical University in southern China. I suggested that we take some Kirlian photographs there and Dr. Fu arranged a meeting.

First we met in traditional Chinese style with several of the leaders of the hospital, each with a cup of tea. They were all curious about the Kirlian camera so I took their Kirlian pictures. They were amazed at the patterns. So was I. Every practicing doctor had a preponderance of streamers! The two who did not regularly see patients had normal pictures. Five doctors with regular contact with cancer patients, and five photographs of streamers. These doctors couldn't all have cancer! It seemed that something about the hospital was drawing electrons from the doctors. And since only those with patient contact showed this depletion, it seemed possible that the cancer patients, or the cancer itself, was draining the doctors!

Now that I had seen the Kirlian images of the doctors, I was eager to see what the cancer patients' Kirlian images looked like. The doctors consented. The first patient put his hand in the bag and I snapped the picture. It was almost blank. We could not discern any pattern of balls or streamers. To increase the brightness of his Kirlian image I put an Infratonic QGM on his foot. The QGM is an infrasonic treatment device which increases a patient's vitality by simulating the emitted Qi from Qigong Masters. After twenty minutes of treatment, I snapped another picture. It worked. I got a bright picture. And, as I had expected, there were no balls, only streamers.

(Left) the Kirlian photograph of a typical cancer patient shows almost no image.

(Right) Applying the Infratonic QGM increases conductivity and reveals the electron deficiency through streamers.

I repeated this test with a total of six patients and five of the six had identical results. A very weak picture at first, and after treatment with the Infratonic QGM, a stronger picture showing streamers. From a scientific point of view, a very low electrical conductivity of the arms was masking a severe electron depletion. The QGM temporarily increased electrical conductivity so that I could get a picture.

Dr. Fu and I felt exhausted afterwards so we took our own pictures, and sure enough, our Kirlian images had shifted toward more streamers as well. How could someone else's cancer have sucked vitality out of us? Modern medicine may come up with some very interesting answers here. I am confident that Qigong will prove to be one of the factors.

Kirlian Photography Supports Yin and Yang

Yin is the magnetic aspect of Qi. It appears in Kirlian photography that electrical conductivity in the hands, wrists and arms indicates the presence of Yin by the brightness of the Kirlian image. A small, broken Kirlian image indicates low, uneven conductivity in the body—low Qi. It appears that Yin is a magnetic substance that is stored in water molecules in the body. It aligns along the electrically conducting structures called meridians that make up the acupuncture system. When our magnetic energy is low, these meridians are weakened and our conductivity is lowered. Our magnetic energy might be replenished from ingesting fresh food, or absorbed directly from Earth's magnetic field. The hand-washing experiment I described in the previous chapter shows that tap water can remove conductivity, and "magnetized" water can add it.

Where the Kirlian image is dominated by streamers, the body has some conductivity (Yin) but is deficient in negative electrical charges (Yang). In this case, a great number of electrons flow toward and into the fingers and few flow back out again. Thus, long streamers indicate a deficiency in Yang. During the Kirlian photographic process, the extra electrons are collected from the air in the form of free ions and ionized oxygen molecules. Ionizers, sold as air purifiers, shoot billions of free electrons into the air. These negative ions appear to be an aspect of Yang.

The cancer patients I photographed were unable to produce streamers until the Qigong machine was used on their feet. Thus, just having negative ions in the air may not be enough to get them to the site where they are needed in cancer patients. Many Chinese doctors insist that the patient's will can extract the vitality from the cancer and return it to the patient's body.

If the image is bright and consists of a balance of both balls and streamers, then it would appear that the person has both the Yin, magnetic aspect and the Yang, electrical aspect. This is consistent with the idea that a balanced combination of both Yin and Yang results in strong Qi.

The Power of Ions

Much research has been done into the effects of negative ions. These ions remove airborne pollen, soot, cigarette smoke, and household odors, either by causing it to collect into large molecules which precipitate to the walls or floor, and second, by activating chemical reactions which make the pollutant inert. This can be a big advantage to those who suffer from asthma or allergies.

Further research indicates that negative ions affect human functioning as well. Negative ions in the blood stream increase the oxygenation of our cells and tissues, and are critical to the functioning of neuropeptides at nerve junctions. Positively charged ions, or a deficiency of negative ions causes a depression of these activities. Future research may reveal that these negative ions are central to life as we know it. For instance, some view negative ions from a metaphysical perspective as critical to the processing of life experiences. From this perspective magnetic flux is viewed as partially completed action, suspended thoughts, or karma. Electrons bring these suspended thoughts into activity, and through the individual's acceptance, to completion. At this point, a positron is released, which annihilates the electron, releasing two photons of light. This light can be viewed as karma brought to completion. This parallels the Chinese view of transmuting Qi to Shen, that will be presented in the next chapter.

The Guy with the Halo

I mentioned that five of the six cancer patients tested out the same, first with a very weak picture, then, after the Qigong machine, with streamers. The sixth patient was exceptional. He had a very bright glow with lots of streamers. After applying the Qigong machine to his foot for 20 minutes, we photographed again and found no balls and no spikes; there was just a big round glow with smooth edges around each finger.

I had seen these halos before, and at first thought that they were accidents. When they continued to show up occasionally, I just ignored them. I could have ignored this cancer patient's halos as well, but I didn't. The halos were apparently a third structure independent of balls and streamers, unaffected by electrical conductivity of the arm. What could they be? What was it about this man's fingers that caused these halos?

I asked the attending doctor, "Is there anything different about this man?"

The doctor was surprised by the question and answered, "He has cancer like the other patients, but he *is* different. He is optimistic. He is not worried about dying. He always has a kind

(Left) One cancer patient who showed great optimism also showed a bright Kirlian image.

(Right) After the Qigong machine, his Kirlian image turned into a halo.

word. Though he is very weak, he is eager to help other patients." This was in stark contrast to most cancer patients who spend much of their time depressed and worried.

To explore this mystery and the big question in my mind, How is the Kirlian image related to Qigong? I set out to measure the Kirlian images of Qigong Masters.

Key Points from Chapter 3:

1. Increasing the magnetic substance in a test subject increases internal electrical conductivity, thus increasing brightness of the Kirlian image. This magnetic substance is similar to the Chinese concept of Yin, the water, the nurturing aspect.

2. A deficiency in electrical substance causes long streamers in the Kirlian image as electrons stream from the film to the finger. Ion generators normalize the Kirlian image, indicating that free negative ions are a key component to Qi. The lack of activity among cancer patients and others who show long streamers indicates that free negative ions are the Yang or active aspect of Qi, often characterized by light or heat.

3. The cancer site has balls without streamers indicating a high concentration of negative ions. The cancer exhibits high activity, as observed in rapid growth. Fingers of the cancer patient show extensive streamers and often, the patient shows very weak activity. Thus, it appears that the cancer draws this Yang aspect of activity from the patient. Research into methods for transferring the Yang from the cancer back to the patient may prove valuable in the ongoing search for effective cancer treatments.

CHAPTER

KIRLIAN IMAGES
OF QIGONG MASTERS

My interest in Kirlian photography started when I saw Kirlian photographs of the hand of a Qigong Master before and during Qi emission. They showed me that this Qigong Master had a much brighter Kirlian image during emission than before. In my early research, I had seen that both cancer patients and healthy people showed brighter pictures when they had more Qi, so I expected to see that Qigong Masters, who spend their days collecting Qi would have bright pictures all the time.

Photographing Masters

Back in 1990, soon after I got my first Kirlian camera, I was taking Kirlian photos at an acupuncture symposium. Master Effie Chow stopped by to get her photo taken. At first, she didn't let on that she was anything but an acupuncturist and a nurse. She had a very bright Kirlian image and was very sensitive to the electrical flow from the camera, feeling it up to her shoulder.

Master Chow invited me to give a presentation in San Francisco for the Qigong Institute. There I took a Kirlian photo of Master Gong, another guest lecturer that day, who at first produced an average picture, then, with practice, produced a most

31

bright and impressive picture. I thought I had it all figured out. Masters must have lots of Qi, and therefore produce big pictures. But I had uncovered no more than a corner of the truth.

Then Dr. Ken Sancier, formerly a senior scientist at Stanford Research Institute, and now Director of Research at the Qigong Institute, suggested that Dr. Fu and I visit Master Zhou Ping-jue who is considered to be one of the most powerful Masters around. He was about 60, short, mostly bald, and very solid. On his wall were pictures of big trucks running over his body.

Before Master Zhou would let me take his Kirlian picture, he demonstrated his Qigong skill by placing a piece of aluminum foil, four inches by ten inches, on a wet paper towel and folding the combination into a thin pad about three by four inches. He then placed the pad on my knee and started emitting Qi, holding his hand six inches above my knee where the pad was resting. Soon it started steaming, and got so hot that I had to pull it away.

After it cooled he rolled it up into a cylinder about two inches long and one-half inch in diameter and rolled it back and forth between his hands. Steam started shooting out of both ends. He handed it to me. It was so hot that I threw it in the air. He caught it and handed it to Dr. Fu. She felt its heat and quickly threw it back to him too. I was impressed. All I could think of was that Master Zhou's Qi field, or a powerful emission of some form of negative ions, had somehow catalyzed the oxidation of aluminum in water, releasing lots of heat. I found this demonstration to be quite remarkable because Master Zhou could cause the aluminum to heat up at will, without touching it.

When I showed Master Zhou the Kirlian camera, he explained that electrical equipment often malfunctioned with him around. He had apparently been stung by Kirlian cameras before. He stuck his hand in the bag and I snapped the picture. The film was blank. Not like the cancer patients who had a spot of light here and there, but totally blank. It was embarrassing to have him fail so miserably. He felt bad because he couldn't make a bright picture. And I felt bad because I had begun to doubt whether I knew what the Kirlian camera was really measuring.

He suggested that I try to take Dr. Fu's picture with him emitting Qi. I took her picture before the emission, and she had a normal picture. Then I took another picture while he was emitting Qi down her arm in order to increase the Qi flowing out her hands. Now this was really embarrassing. Her image shrank. It looked as if he had stolen her Qi. This was quite a puzzle. A Master who performs one of the most impressive demonstrations I had seen, and who by many accounts is quite a capable Master, has no Kirlian image at all and makes other people's images shrink!

I packed up my Kirlian camera and headed to China, where I knew I would find more Masters to test. I tested two of Master Wan's students, both powerful Qigong therapists. Both came up blank. They went to get Master Wan, and his photo came out blank as well. This time we had a high degree of consistency. Three Masters in a row and three blanks.

We took my picture again to make sure the camera was working, and indeed, my image was normal. Unfortunately I had done it again, by showing that my picture was bigger than theirs, I implied that these Qigong Masters were wimps. They tried and tried again to make bright pictures, but they couldn't. I tried to emphasize the scientific importance of this event, but I could tell that their tolerance was wearing thin.

My mind was spinning for a way out of this one when I had a stroke of genius. I asked them if they could stop doing the Master thing for a few minutes. Could they just be normal people like me? They thought for a few minutes. I could see that they were straining their brains trying to get out of the Qigong state. On their next try, Master Wan and Little Zhou managed to make pictures that looked quite a bit like mine. Little Huang, a cute 100-pounder, tried and tried, but couldn't stop being a Qigong Master, the poor thing. Little Zhou, the student who always seemed to be best at everything, tried again, and this time, produced a really bright picture. That night, I couldn't sleep, plagued with the question "Why?" What had I just learned?

Kirlian Photos at a Qigong Lecture

Dr. Sancier conducted some interesting research with one of my Kirlian cameras. Master Yan Xin, probably the most famous Qigong Master in the world, was lecturing in the local area, and Dr. Sancier set up a study to determine the effects of Qigong meditation and Qigong lectures on participants. Master Yan lectures to crowds as large as 20,000 people, inducing a Qi field that engulfs the entire crowd, with many claiming miraculous healings just from sitting in the audience. Numerous scientific studies have been done in which Master Yan emitted Qi, such as mending broken bones at a distance, altering growth rates of plants and bacteria, and changing the decay rate of radioactive substances. The results of these studies are most impressive.

Dr. Sancier tested a group of Qigong practitioners before and after doing Qigong exercises, and found that the brightness of their Kirlian images doubled as a result of their efforts to collect Qi. His results indicate that Qigong exercises increase the amount of electrically conductive Qi in the body.

Then Dr. Sancier tested a group attending Master Yan's lecture. During the lecture the audience was encouraged to do

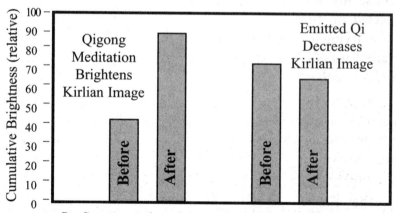

Dr. Sancier conducted research which indicates that, while Qigong meditation increased the brightness of the Kirlian image, the effect of Masters emitting Qi was variable. Overall, Master Yan's emitted Qi caused a slight average decrease in brightness.

Qigong exercises. At the same time, everyone was exposed to the Qi field of Master Yan. The result was a wash. Some pictures of audience members were much brighter after the exercises and lecture. Other audience members had pictures which were much smaller. On the average, the brightness of Kirlian images was slightly less after the lecture.

Dr. Sancier sent the Kirlian camera back with a copy of his results. I can understand his disappointment, much like that of early Kirlian researchers. Some images got bigger, some got smaller. There seemed to be no consistency. Had Dr. Sancier been measuring two variables which cancelled each other out? If the first variable was electrical conductivity, what was this second variable?

My Qigong Treatment

Still in China, I puzzled at why a Master often has no Kirlian image and why he sometimes makes others' images go away. Dr. Fu asked three of Master Wan's assistants, Little Zhou, Little Huang, and Little Cui, to give me a Qigong treatment to see how it would affect my Qi. At that point I was a challenging test subject because my Kirlian image was always about the same. As expected, my first picture was "normal for Richard." Masters are always trying to get me to feel something by aiming Qi at me in different ways. They are nearly always disappointed because I never feel enough that I can be sure it wasn't my imagination.

As they worked on me (without touching me) I felt nothing. This was also "normal for Richard." Half-way through the treatment a second picture was taken which showed that my image had gotten weaker. They felt a little guilty, as if they were stealing my energy or something.

They finished up and photographed my fingers again. My Kirlian image had virtually disappeared. This was quite a puzzle because it implied that, whatever they did to be in the Qigong state, they had done to me also.

35

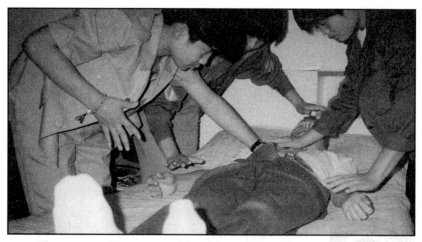

Three young Qigong practitioners treat the author with emitted Qi.

Have you ever been sitting there puzzled about something, when suddenly the answer slams you in the face? It wasn't quite like that. It got me in the hand. Something was different about the feel of taking the Kirlian picture. They snapped another one. I heard the characteristic zzaaaapp as the 30,000 volts was released and the current flowed between me and the film. But I heard it from the wrong hand! I pushed the button again, and, indeed, the zapping sound occurred where I was holding the ground electrode. I pushed the button a third time, and heard the zap, but felt and heard nothing from my hand that was placed on the film. The noise and electrical sensation came from the hand holding the electrode.

Whereas the biggest electrical resistance is normally between my hand and the film, this time, the electrical resistance was between the fingers of my other hand and the ground electrode. Electrons were somehow moving from my fingers to the film without high voltage and without ionizing air!

This could only mean that the Qi field around my hand produced by the Masters somehow allowed electricity to flow through the air more easily. The Qi field makes air into a bit of a superconductor, or else somehow allows electrons to tunnel through hyperspace. Either way, *this clearly is a major alteration in the laws of physics — an alteration induced by consciousness.*

When a Master waves his hands over a patient's painful joint or a broken bone, the healing process appears to accelerate. Perhaps the Master's Qi field affects human tissue the way it affects air, giving it superconducting attributes, restarting a healing process which had stalled, allowing chemical reactions to happen far more easily than normal. This may be how Master Zhou catalyzed the oxidation of aluminum foil.

(Left) My Kirlian image shows numerous balls and streamers before the Qigong treatment.

(Right) After the treatment, my image softened, transforming into a halo.

Studying my Kirlian images I noticed that, while the brightness of my Kirlian images had decreased, the texture had also changed. Instead of a collection of balls and streamers, they looked more like halos, round and smooth. These young Masters had turned my balls and streamers into a halo. My mood shifted as well. I felt more optimistic.

The shift I saw in my Kirlian image from balls and streamers toward the halos was a measure of some sort of energy transformation in me. My hand had gained some of the superconducting attributes that Masters have. *The halo appearing in the image of my hand was a transition state between the balls and streamers of a normal healthy person, and the blank pictures of Qigong Masters.*

I managed to verify that the halo is the transition state one day when my brother John was visiting. We had the Kirlian camera out and I challenged Dr. Fu to make John's picture bigger. She started emitting Qi, and I started taking pictures. Within five minutes, his Kirlian image gradually shifted from balls and streamers into a halo, then faded away. She failed in her assigned task, but filled in a missing link. The halo replaces the balls and streamers as the Qi field of the Master, or of the patient receiving treatment, increases. Then the halo itself fades away as the Qi field gets even stronger.

Understanding the Halo

I puzzled about the halo and what could cause it to appear. As electrons shoot from the film in the Kirlian camera to the fingers, they form long streamers. And as electrons flow back from the fingers to the film, they pool, forming balls. The halo was something different. It was a uniform glow. Whereas most people's Kirlian photos show balls and streamers to some extent, only a few people have the halo. What could this mean? And how does the halo progress to no picture at all even though the flow of current increases?

We can understand that electrons cascading toward the finger can create the streamers, and that electrons spreading from the finger can create the balls, but what sort of phenomenon can create the halo? How can electrons possibly travel through the air from the finger of a person to the film without ionizing the air? This is certainly possible. It is accomplished in ion generators where electrons are forced to the tip of a sharp needle with minimum voltage drop and with no ionization of the air.

However, it appears that electrons are conducted between the finger of a Qigong Master and the film in a different way, perhaps like a superconductor. They leave the finger silently, without an ionizing jolt. And they flow smoothly outward to their destination. They glow as they move and produce white light instead of the blue light characteristic of the ionized air of the Corona discharge. As this field becomes even more conductive electrons flow even easier, without producing any light at all.

Perhaps Shen, the spirit, the substance of mind or will, said to be produced during Qigong exercises, changes the electrically conductive nature of air, or changes the nature of electrons as they travel through air. It is as if, without Shen, electrons travel like particles, and with Shen, they travel like waves. This parallels quantum mechanics: In normal Kirlian photos, electrons act like particles, traveling in straight lines, ionizing air molecules as they move from the film to the finger, yet in photos of Masters, electrons act as waves, disappearing from the film and reappearing at the finger without ionizing any air molecules in their path.

In his book **The Holographic Universe**, Michael Talbot reports research in which advanced meditators are placed in a shielded room. Before they enter the room, special sensors measure cosmic rays passing through the room, indicating that the normal cosmic radiation from the sun and deep space is penetrating through the shielding, as it is known to penetrate all known matter. But once the meditators enter and start to meditate, cosmic radiation is no longer measured passing through the room!

Somehow, these meditators have the same effect on cosmic radiation as Qigong Masters have on electrons during Kirlian photography. They act like waves instead of particles, somehow going around instead of creating ionizing radiation on the way through. This suggests that those who do regular Qigong or other types of meditation may somehow protect their chromosomes from the medical dangers of cosmic radiation and other ionizing radiation. I was on to something very interesting!

I had compared the conductivity of the body to Yin and the electrical charge of the body to Yang. Perhaps there was more to the analogy. Was the halo also identified in the Chinese system of collecting Qi? As I pondered, I kept coming back to what Chinese doctors view as the basic building blocks of human life.

Jing is the Qi of the kidneys which acts as a catalyst to grasp the earth Qi from the digestive system (Yin) and combine it with the heaven Qi from the lungs (Yang). This process involves deep slow breathing and mentally directing this collection process, usually focusing the resulting Qi to be stored in the lower abdomen. Qi, the result of this combination of Yin and Yang, closely parallels the "vital energy" of the western philosophies.

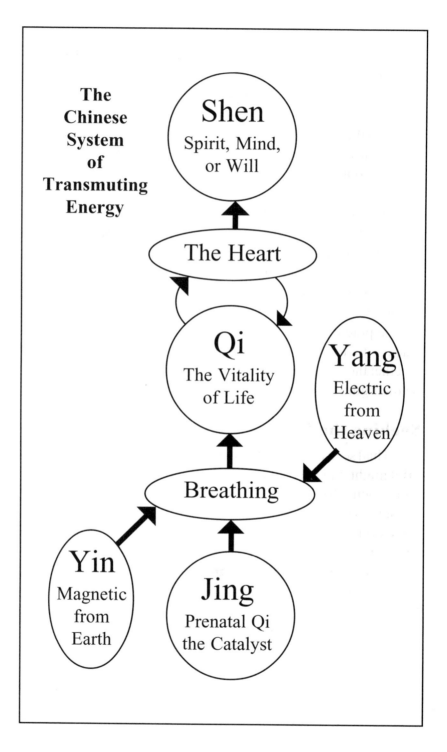

The Chinese System of Transmuting Energy

Shen
Spirit, Mind, or Will

The Heart

Qi
The Vitality of Life

Yang
Electric from Heaven

Breathing

Yin
Magnetic from Earth

Jing
Prenatal Qi the Catalyst

A second step in vitalizing the body, according to the Chinese, is to circulate this accumulated Qi up the back, around the head and down the front line of the body, then to store it again in the lower abdomen. Circulating the Qi around this "microcosmic orbit" activates the heart to manufacture Shen. This Shen is stored in the pineal and pituitary glands in the brain. A combination of the accumulation of Shen and the intent of the individual brings about the opening of the third eye and the awakening of higher levels of human functioning. This might explain how Qigong doctors are able to see into others for diagnosis and how clairvoyants are able to see the future.

This process of first manufacturing Qi, and then converting it to Shen, is the most fundamental technology of Qigong science. The problem is that students of many pathways of spiritual development are not sure whether they are really doing it right. People practice for decades before they "get it." They go from teacher to teacher, trying to understand, but never knowing whether they understood. It appears that Kirlian photography can be their guide.

Seeking the Halo

I had seen enough to believe that circulating the microcosmic orbit might, indeed create Shen and result in the appearance of the halo around the finger. Shen Gong is a form of Qigong in which the purpose is principally to cultivate Shen. To test this hypothesis I sat every day, for a week, for two to three hours doing the Microcosmic Orbit exercise. I directed energy up my spine and down the front of my body, focusing particularly on smiling, feeling love, and feeling the Qi flow through my heart on its way to my lower abdomen.

I, of course, took pictures every few minutes at first, hoping that I would see a shift. At the end of the first day, I thought I saw some shortening of the streamers. By the end of the second day a slight glow was appearing. By the end of the week I had created pretty good halos around my index and middle fingers. My ring and baby fingers had little halos, but clearly needed more work. I felt great love and contentment that week as well. If you wish to change the course of your life, I believe you can start to do so within a week if your meditation is intensive.

Experiment with Meditators

To test this theory of mine, I asked participants at a Life Enhancement Workshop in Washington, D.C. about health, happiness, and their practice of prayer and meditation. I hoped to get an idea of what practices were related to halo formation. The correlations were clear:

The three participants who practiced prayer or meditation at least three hours per day all showed well-developed halos. About a third of those who practiced more than fifteen minutes per day of prayer or meditation had signs of halo development. Almost none of those who did little or no prayer or meditation had halo development. This supported my hypothesis that prayer and meditation achieve the process of transmuting Qi to Shen. Few of these people had been exposed to Qigong before, so I would add that, regardless of religion or philosophy, loving contemplative activity enhances the halo. I believe this is a measure of Shen, of mind.

I should add that another correlation appeared in this study. Those who had a regular practice of prayer or meditation indicated that they had far fewer health problems than those who did not. Some well-controlled research studies have shown these same benefits to the health of those who pray or meditate regularly. Conclusion: Not only does prayer or meditation make for pretty pictures, but it's good for you too!

Photographing More Masters

I had the opportunity to photograph many Qigong Masters at the Third World Conference for Medical Qigong Exchange in Beijing, and learned that the phenomenon of a completely black Kirlian image can be as a result of very high Shen, very low Qi, or a combination of both. Those with moderate Shen and strong Qi will produce the halos around the fingers whereas those with moderate Shen and weak Qi (those who just gave an emitted Qi or massage treatment), will have no image or just a faint glow. However, those with very strong Shen will probably have no image at all regardless of how strong their Qi is. So, if you find someone with no picture at all, here are some guidelines to sort out those with low Qi and those with high Shen:

1) Make sure they were touching the film and touching the ground electrode. If the circuit is not complete, no electricity will be produced, resulting in a blank picture.

2) If they felt a strong electrical flow, or jumped when the picture was taken, this is a sign that they are very sensitive and that high current flowed, indicating that they have high Shen.

3) To strengthen a very weak image, you might give this person a five-minute shoulder massage to increase the electrical conductivity of the hands, then take the photo again. Ten minutes with the Qigong Machine under the foot should work too. If weak Qi caused the blank photo, you should now see a stronger image. Alternatively, you might instruct them to do Qigong breathing to collect more Qi. This usually increases brightness of weak images as well.

4) Take their picture first with four fingers on the film, then make a second exposure of just the index finger on the same film. All of the current from the camera will flow through this one finger, making a much brighter image. This will allow you to see the nature of the image, whether balls and streamers or the halo. If that one finger shows a halo, strong Shen is the dominant factor. On the other hand, if you see the balls and streamers, Qi dominates and this person is weak in Qi. If you can't get an image with a single finger and your test subject is not suffering from severe exhaustion, he probably has very strong Shen. I suggest you bow.

5) If you suspect someone is a Master, ask him to try to become a normal person, to withdraw the Shen, or awareness from his fingers. A capable Master will probably eventually learn to

make many kinds of images at will, though I have found two documented powerful masters who had such strong Shen (and perhaps low Qi) that they could not produce an image no matter what they did. Both had demonstrated the catalytic heating of aluminum before being photographed, which, perhaps, exhausted their Qi.

Another factor you should be aware of is that one powerful Master in the room may affect others, so if you find that several people have blank pictures, consider the possibility that a person with high Shen is allowing her field to penetrate others in the room. You might try asking her to "keep her Qi to herself" for a few minutes or to step outside for a few minutes. Be sure to invite her back in, though, because people with strong Shen can be very good luck.

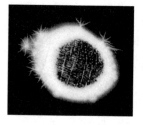

This Kirlian image is of a Qigong Master who, after many blank pictures and many attempts to "be a normal person," finally produced an image which reflects very strong Shen. This isn't normal.

On the other hand, some people whose Kirlian images showed no signs of Shen showed impressive paranormal skills, and others who showed strong Shen appeared to be out to serve only themselves. So while the halo around the finger is an important new measure of human development, and can indicate special healing ability, it is not an absolute measure.

Kirlian photography appears to quantify aspects of human health and development of consciousness which have, until now, been ignored by western science. These aspects, identified in traditional Chinese medicine, are Yin, Yang, Qi, and Shen. Cultivating and balancing these vital substances is the purpose of Qigong and the basis of Chinese medicine. Kirlian photography is thus, a window into successful practice of Qigong and into understanding traditional Chinese medicine.

*Master Duan emits Qi into a glass of wine to reduce
the intoxicating effects of alcohol. This 89 year-old
Master, a Catholic, insists that Qi is a part of all of
the world's great philosophies and religions.*

Key Points from Chapter 4:

1. The halo appears in the Kirlian image of healers, and specifically around those fingers they use for healing. Shen Gong, the Chinese exercise of circulating Qi up the back and down the front of the body and through the heart, strengthens the halo, as do extensive bodywork, prayer, and meditation.

2. Shen, as it increases, gradually converts the rough Kirlian image of balls and streamers to a smooth symmetrical one. This appears to be a superconducting or fourth-dimensional phenomenon by which electrons travel between the finger and the film without ionizing air, a great opportunity for scientists to study the physics of paranormal abilities.

3. As healers give treatments, their Kirlian image gets much weaker, or disappears entirely, as the Qi diminishes but the Shen remains strong. Electrical flow through the camera remains high, but air is no longer ionized and light is not produced.

AIRPLANES DRAIN QI

Long airplane rides are known to be exhausting. If the Kirlian camera really measures vitality, then, I reasoned, we should be able to observe changes in the Kirlian image during airplane flights. Further, it seemed that free ion generators and magnets might be used to supplement the electromagnetic fields available in the aircraft to normalize the Kirlian image.

I do extensive flying, mostly between Los Angeles and China, and know that it's tiring work. It has always been a puzzle to me why it is so much harder to sit on an airplane for 12 hours than it is to sit in a chair at home for the same amount of time. Dr. Fu is very disturbed by air travel. Whenever she flies, she feels dizziness, spontaneous perspiration, rising heat, headaches, nausea . . . She doesn't like to fly. Her symptoms are common for people who suffer from what, in Chinese medicine, is called Yin deficiency, which might be equivalent to deficiency of coolness, fluids, or, in Kirlian terms, deficiency in magnetic substance.

My hypothesis was that an airplane, because of its position above the earth and its metal structure, decreases the availability of both electrical and magnetic substance. First, the earth's magnetic field at 30,000 feet is much weaker than at ground level. And second, because the Earth has a strong electrical field drawing electrons to the ground, there should be fewer available electrons, or free negative ions, available. Actually, airplanes have other problems as well. The air-conditioning system will tend to strip out any remaining free ions before delivering the air to the cabin, and the hundreds of people on board will be competing for the few remaining negative ions.

You may argue that dryness of the air and concentration of airborne bacteria are important factors. You are probably right. Actually, I have experimented with water, and have found that, by drinking lots of water and by avoiding substances like tea and alcohol which upset water balance within the body, I have been more comfortable on long flights. And research shows that negative ion supplementation helps to kill airborne bacteria. And people who are deficient in the Yin and Yang aspects of Qi appear to be more succeptible to bacteria. So the problem of airborne bacteria on airplanes may be greatly increased by ion deficiency.

For one flight, I set aside all these other variables and focused on the electrical and magnetic aspects of Qi that might be affecting air flight comfort. I gathered up my equipment: a Kirlian camera, several battery-powered ion generators, and assorted magnets, and headed to the airport for a trip to China with 25 doctors. We were off to Beijing to study Qigong, so it was a perfect opportunity for this group to participate in an air travel experiment.

Air Travel Increases Streamer Formation

At the airport, before the flight, I took everyone's Kirlian photograph. A few hours into the flight, I took their Kirlian photos again and discovered that two thirds of them showed a significant increase in streamer formation, indicating that they were, indeed, being depleted of free negative ions by the airplane flight. Those who showed no change didn't have any streamer formation in the initial photos. By the end of the flight nearly everyone had some streamer formation, and many had dramatic changes.

Those who started with small images and showed rapid change to streamer formation were the most uncomfortable on the flight, with symptoms of Yin deficiency similar to what Dr. Fu experienced. I distributed the free ion generators to those who showed the greatest streamer formation and experienced the most symptoms of discomfort. Within an hour, their streamer formation had decreased significantly.

One traveler was miserable with Yin deficiency symptoms, so I started her out with magnets, thinking that a stronger magnetic field would strengthen her Yin and increase her comfort. An hour later, she was still uncomfortable. Her Kirlian image had brightened considerably but her streamers had grown as well, suggesting that she had gained Yin or magnetic substance, but continued to lose the Yang aspect of negative ions.

I traded the magnets for an ion generator, and came back after another hour. She was still uncomfortable, but her Kirlian image had changed toward substantially reduced streamer formation. I then let her wear both the magnets and ion generator. After another hour, she felt better and stayed that way throughout the flight.

| *Before the flight. No streamers* | *Two hours into the flight. More streamers* | *After 1/2 hour of using magnets. Much brighter* | *After 1/2 hour of magnets and ions. Strong Qi* |

*Fellow traveler with Yin deficiency
shows changes through the flight to China.*

Flight Attendants' Burnout

Two flight attendants became curious and wound up volunteering to participate in the experiment as well. Their Kirlian photos showed very weak images and only streamers. They then wore the free ion generator around their necks for ten minutes. Subsequent photos showed much brighter images. It appears that,

49

through experience of many flights, they had somehow learned to protect themselves from the airplane's ion drain. When they started wearing the free ion generator their bodies figured out that there was an available supply of free ions and the pores of their skin opened up to absorb as many ions as possible.

(Left) Flight attendant after four hours of flying.

(Right) Same flight attendant after using free ion generator for ten minutes.

Like nurses, flight attendants often burn out. The drain of negative ions may be part of the cause of burnout. In the above experiment, we may have observed how flight attendants who don't burn out have learned to survive in an adverse environmental situation.

One last thing we learned on that flight was that a magnet placed with its negative pole facing the base of the neck at C-7 removes the tension that builds up there during airplane flights. After this discovery, I was amazed to find that this tension in the neck is the principal physical complaint of most air travelers.

Recovery From Plane Flights

After our 12 hour flight and arrival to the hotel, I fell fast asleep at about 9 p.m. Beijing time. I woke up again at 1 a.m., wide awake. So I took my Kirlian photograph. It showed even more streamers than it had on the plane. I got out an ion generator and breathed the negative ions for about 30 minutes. I then took my photo again and was much relieved to see that the streamers had diminished considerably. I fell back to sleep with the ion generator still running.

The next morning, I went walking in Fragrant Hill Forest, which surrounds our hotel, and did deep breathing as I walked, enjoying the fresh morning air. When I returned, my Kirlian picture was even smoother and brighter. It appears that natural forests are even better than free ion generators.

| *Before the flight* | *1 a.m. after the flight* | *After 1/2 hour of ion generator* | *After a walk in the park* |

Ion generators and natural ions from trees contribute to recovery from long plane flights.

For all those readers who dream of flying in comfort by flying First Class, be aware that First Class also drains Qi. It is not like a seat in the forest. A few magnets and a free ion generator may be a better investment than a First Class upgrade.

BIOELECTRIC VITALITY

> ## *Key Points from Chapter 5:*
>
> **1.** There are two principal reasons for discomfort in air travel. The first is the low availability of free ions in the cabin which results in Yang deficiency conditions. The second is the low availability of earth's magnetic field which aggravates conditions of Yin deficiency.
>
> **2.** The deficiency conditions caused by low electric and magnetic fields on airplanes can cause lingering symptoms and succeptibility to airborne bacteria and viruses.
>
> **3.** A free ion generator with magnetic supplementation, worn around the air passenger's neck relieves both the electron deficiency and the magnetic deficiency by creating a local environment similar to that in nature. This can substantially increase passenger comfort and protect health.

CHAPTER 6

THE ATMOSPHERE IS ELECTRIC

While flying over the Pacific, I have had much time to experiment with the atmosphere of airplanes. I had thought that, since there is such a strong electrical gradient pulling negative charges toward Earth, there might be a shortage of negative charges at high elevations, yet negative ion generators seemed to work just as well up there as at ground level. Could our entire atmosphere be negatively charged?

When my son, Tomu, stares out the window, he asks the inevitable question, "Why is the sky blue?" I launch into the common explanation of how Earth's atmosphere scatters the light from the sun. I conclude with the observation that as the sun sets, the sky often turns red because of this scattering.

While Tomu accepts my answer, I can not. Diffraction of Earth's atmosphere can account for sunrise and sunset, but not for light coming straight in through the atmosphere at noon. And photographs taken from space show Earth as a planet surrounded by a blue atmosphere.

In a dark room, the Kirlian discharge that is recorded on film appears as the same color blue as the sky. The corona discharge around needles of powerful negative ion generators also appears blue. I know from the odor that this is due to ozone, which is defined as: "An unstable, *pale-blue* gas with a penetrating odor." And right below that in the dictionary was another term: "ozonosphere, the atmospheric layer extending from a height of about 6 miles to about 30 miles in which there is an appreciable concentration of ozone."

The sky is blue because of the 25-mile thick layer of light blue ozone! This leads directly to a very important finding which may come as a surprise to many scientists reading this book: Since ozone is a negatively charged gas which carries an extra electron, and since there is so much ozone in the atmosphere, this implies that Earth is flooded with excess electrons, and carries a very strong negative charge! I had assumed that Earth, overall, would carry a neutral charge. As you will see, this negative charge has a strong influence on life as we know it.

One evening, Tomu and I settled down with a children's book in which a magic school bus travels through the solar system. When we were blasting off, we looked back at a beautiful picture of Earth, surrounded by dazzling blue atmosphere. I tried to explain how ozone created that beauty, but Tomu would not let me gloat. He immediately pointed to a picture of Mars and asked, "Why is Mars red?"

I gave him the obvious answer, "Because it doesn't have ozone." Again he accepted my answer but I was left with a feeling of emptiness. Mars and Earth are so similar in size and position relative to the Sun. Why is one blue and one red? Why do we have all this wonderful ozone protecting us while Mars has a skimpy red atmosphere?

Then it hit me like a bolt of lightning. The growing hole in the ozone layer! Earth is losing its electrons! If we lose our net-negative charge we will lose our ozone layer. With a zero charge, or net-positive charge, the ozone will be replaced by an ionized layer of nitrous oxide! This is the "photochemical smog" which turned the air of our big cities red in the 60's and 70's.

On a recent trip to Australia, I was warned to stay out of the sun because of very high ultraviolet exposure. This was caused by the hole in the ozone layer which extends over Australia during the antarctic summer. As I looked up at the sky, I saw deep grey with a tint of blue, very different from the brilliant deep blue and powder blue skies seen from the United States.

To the extent that air pollution and fluorocarbons are responsible for depleting the ozone layer, we might be able to offset this loss and clean up our cities through mandated programs of ionization. A small room ionizer consuming just a few watts can produce a trillion ions per second. Low cost commercial grade ionizers on airplane wings and cars could flood the skies, particularly our big cities, with negative ions, cleaning up air pollution and replenishing the ozone layer.

Qi Field of the Earth

The Earth, like all creations, has both a Yin, supportive aspect, its magnetic field, and a Yang, electrical aspect, its ozone layer. They are interdependent.

Earth, as it rotates about its axis carries its strong negative electrical charge around this rotation. This creates much of the powerful magnetic field that surrounds Earth and protects its atmosphere from the powerful solar wind. This magnetic field, in turn, creates a strong electrical field which pulls electrons powerfully to Earth's surface. Lightning is an illustration of this powerful field as rapidly rising air carried electrons from Earth upward against this voltage. When enough electrons collected in the clouds, their voltage differential built up, reaching millions of volts. These electrons shoot back to Earth as lightning bolts.

These streamers are electrically quite similar to the streamers in Kirlian photographs which occur when electrons shoot from the negatively charged film to the finger.

This Kirlian photo shows electrons shooting to the finger.

Earth is losing its electrons as indicated by growing holes in the ozone layer. Along with the weakening electrical charge, Earth's magnetic field is getting weaker. According to John White in **Pole Shift** the magnetic field of Earth is about half as strong as it was 2500 years ago. He writes, "If this trend continues, the field will disappear altogether—perhaps as soon as early in the next century."

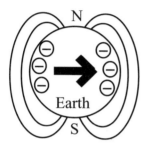

A negative charge at the surface of the Earth rotates about Earth's axis, creating a powerful magnetic field. This rotating magnetic field creates an electric field which holds the electrons close to Earth, sustaining this electromagnetic structure and protecting earth's atmosphere.

This is Earth's Qi body!

In the book **Cross Currents**, Robert O. Becker, M.D. writes, "The evidence is that we are in the initial stages of a reversal. The average strength of the natural magnetic field has been gradually declining for the last several decades."

With the decrease in magnetic field strength and growing holes in the ozone layer, Earth may presently be unstable. Any major geological disturbance which sends ions into the upper atmosphere, establishing a hole in the magnetic field or an electrically conductive pathway for electrons to escape *en masse*, might set off a sudden loss of electrons and bring about a collapse of our remaining magnetic field. Major volcanic eruptions might do this. A close encounter with a comet, or a meteorite which disrupts our magnetic field is a more likely cause.

Another likely cause is a nuclear bomb, which will cause ionized air to convect high into the stratosphere. A detonation at the poles is much more likely to trigger the collapse than one at the equator because of the fragile structure of the magnetosphere at the poles. The shape of the magnetosphere at the poles is shown on the next page. The magnetic duct seals in the ions at the poles.

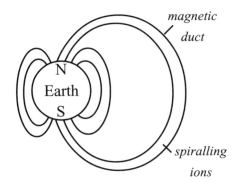

Earth's magnetic field is weakest at the poles due to the fragile nature of the magnetic duct which connects north and south poles.

If the magnetic field does collapse, it will no longer hold electrons at Earth's surface, and they will accelerate away through ionic repulsion of like charges. From my studies of Kirlian photography I learned what the electron exodus would look like. They will not leave like the streamers. They will leave like ball lightning, just as electrons leave fingers which are positively charged through Kirlian photography.

This Kirlian photo shows balls coming off of a finger, indicating an abundance of electrons. This is similar to the ball lightning that will be observed if electrons start leaving the earth rapidly.

Dr. Becker writes, "We now know that the Earth's geomagnetic field has reversed itself many times in the past, with the North and South poles trading places. Humanity has never been through such a "magnetic reversal" but other species have, with dire results." He notes that these reversals correlate with times during which many different animal species became extinct. It appears that, while pole shifts happen, and can cause destruction, Earth has apparently passed through many of these pole shifts without turning into a lifeless planet like Mars.

So, what's going to happen? Psychics have been predicting earthquakes, pole shifts, volcanic eruptions, flooding, and strange weather patterns. If psychics are so smart, how come California isn't at the bottom of the ocean? One might ask what happened to all the floods and earthquakes which were supposed to devastate our world? We've had some, but nothing compared to what was predicted.

Uncertain Prophesies

As I examined the millennium prophesies, I found that, while many of the small prophesies have come true, from an end-of-the-world perspective they have been a big letdown.

Let me give you an example. When we were in southern China in November of 1995, Dr. Fu went hunting the streets of Guigang, her home town, in search of herbalists, folks from the countryside who have spent their lives hiking in the hills, collecting rare and powerful herbs. She found a group of older countryside folks with their wares displayed. She spent several days discussing their rare herbs with them, learning how to prepare and apply them. Most patients can be treated with the more common herbs, so these rare, expensive herbs must be used only for difficult cases.

Dr. Fu pointed out a brightness in the eyes of one of the herbalists, and announced that he was a very wise Qigong Master in disguise. She invited him to dinner at our hotel, and when he arrived, I didn't recognize him because while he was still wearing his old, worn-out peasant's clothes, to get dressed up, he had shaved away his scruffy beard that afternoon. If you had seen him in the street of any major American city, you would probably guess that he was a homeless vagrant. He explained that he would never again come to the hotel, but that it was so kind of her to invite him that he accepted this time. He did not want to be in any way associated with the hotel. He wanted to appear as an average person.

Later Dr. Fu drove over to his house and looked for him. She learned that his house wasn't his. It had been loaned to him for as long as he wanted it by a neighbor who thought he was good luck. Afterwards, the herbalist scolded Dr. Fu for driving her car in front of his house. He didn't want the notoriety that a shiny new car would draw to him. That's when she tried to catch him. I don't know what she said, but it was something like, "I know you are a powerful Master. I want you to accept me as your student and teach me things." While not admitting anything, he did offer to tell her something of importance if she stopped by his house the next morning.

"You would do well to leave China soon," he stated. "If you stay in southern China through the end of the year, you may not be able to leave. All forms of transportation may be shut down. This will probably be a small conflict, a sign of bigger things brewing." He continued with a second prophesy. "In March, there will be a larger conflict which will put China against America. Many will die, but this will be nothing compared to what is to come. Nine out of ten able-bodied Chinese men will be dead by the end of the century!"

Dr. Fu discusses herbal treatment methods with an herbalist from the countryside (center) who forecast death and destruction. Thus far, his forecasts have been accurate as far as timing, but way off with regard to the magnitude of events.

As I told Dr. Fu, these predictions certainly added spice to our trip. I was tired of China and ready to head home anyway, so I agreed that we should follow his advice, but Dr. Fu wasn't done yet. She wanted to learn more about herbs and to search out Qigong Masters. She called up another Qigong Master who examined the future and assured us that our future looked clear if we chose to continue our stay and study with him. We stayed.

Within a few days we discovered that the military was on red alert in our province, just a few hundred miles from the Vietnam border. The North Vietnamese military had conducted a few military exercises, like dropping a bomb on China's side of the border, and was amassing troops as if to invade. China was preparing a pre-emptive strike, to knock the heart out of the Vietnamese' nasty intentions, to swat the mosquito. One high level government economic leader came to visit, and while sitting in our hotel room, explained that he was also on red alert, and like all other important government officials, he was carrying a gun. When he pulled the gun out of his pocket and set it down on my bed, the emergency seemed very real.

The next day, Dr. Fu's brother called the airport to confirm our reservations home. His friend who worked at the airport told him that they were considering stopping all civilian flights and temporarily converting the airport to military operations only, but had not yet done so.

This was pretty good evidence that the herbalist had provided us with a solid future prediction, but, to our relief, the emergency faded away from there.

Two months later, after we had returned to the USA, the news started about China shooting rockets at Taiwan. Then, by March, the United States was sending warships to defend Taiwan in the event of an invasion from the mainland. Again we watched. It turned into a battle to intimidate voters in the Taiwanese election for president, and, like the first event, faded away after the election.

The timing of the prophesy was again right: The huge potential for international conflict was there. But, again, the event faded away without so much as a single casualty. This herbalist correctly foretold future events, but may have misstated their consequences. Perhaps not. Maybe this battle was fought in the minds and hearts of billions around the world. More knew about the Taiwan conflict and were praying for a peaceful resolution than were even alive during World War II. As in the Tiananmen Square conflict, perhaps the world fought most of this battle in their minds and hearts, avoiding massive destruction.

The herbalist predicted a third, even larger, conflict before the turn of the millennium. Based on the first two correctly timed "nonevents," we can expect an even bigger nonevent. The biggest world conflict remaining lies in the Mideast and involves the major world religions, and highly political Mideastern oil. A conflict over controlling oil, a bomb in the wrong place, or an attempt to rebuild the Temple on the Mount in Jerusalem would be met with very strong world polarization, and could easily lead to a world war. And nuclear weapons could shatter Earth's Qi field. However, if such a conflict arises and is settled with minimal fighting, then our herbalist will have been right on timing and wrong on magnitude of destruction every time.

This is just one of many examples of prophesy for disaster which never quite materialized. Predictions of disaster are usually provided with the advice "This is a possible future, but does not need to happen. Mind is the builder. Love and prayer can alter the course of events."

The most common future prediction, and one of the few that psychics agree on, is that a pole shift is coming within the next decade. While they used to speak of the pole shift in terms of disaster, they have recently become very positive. "The shift is from darkness into light, from material into spiritual." Is this pole shift somehow associated with the spiritual transition into the New Age?

And what about all the predictions of disaster? Clairvoyants seem to be saying, "Yes there will be earth changes, but their magnitude depends upon humanity. And, yes, people will die, but these changes are not so much about death, but about new life and the birth of new civilizations. We should be preparing people to relocate and to restructure all aspects of their lives. It is a blessing to be on Earth during this initiation." All this optimism, and all this proposed change, are simply too much for a scientist to swallow.

Can We Really Alter our Future with Qigong?

It is a challenge to try to stand between science and revelation, knowing that scientific discoveries are changing the world at breakneck speed, and that, apparently, our collective human will is changing things so quickly that most psychics have given up on making major predictions.

Here are a few things which appear to be going on, that may be influencing the prospects of a pole shift by altering the electromagnetic balance of Earth: Qigong Masters teach us to do Qigong breathing by drawing Qi from the heavens downward into our lower abdomens. We are to open the tops of our heads and draw Yang energy from the heavens, from the sun and the stars. Many Qigong teachers insist that we can do just this.

If this is possible, it would seem to involve the drawing of electrons into Earth's Qi field. This would mean that, if enough of us spend enough time collecting universal electrons, we will be increasing the negative charge of the earth, thickening the ozone layer, and strengthening Earth's magnetic field in the process. I recently met with scientists from China's Space Medico-Engineering Institute who are looking into this. If Earth's survival is indeed in our hands (or abdomens) don't you think we should find out about it?

Here's another possibility. The process of transmutation of Qi into Shen, as described in ancient Chinese literature and in Chapter 4 of this book, apparently involves transmuting magnetic material and electrons into a fourth dimensional substance, Shen.

Kirlian research suggests that this transmutation is measurable. Since we know that millions of people are now practicing this process, and we observe that Earth's electric and magnetic fields are diminishing, perhaps this process of producing Shen is a principal cause of this depletion. Could our acts of love, prayer, and meditation be causing holes in the ozone layer? Absurd, I agree. Nonetheless, scientists are investigating, just in case.

Here's a third possibility, consistent with some of the more far-out new-age predictions. Perhaps, in our efforts to produce lots of Shen through our love and prayers, we are creating enough Shen, the substance of consciousness, that Earth, itself, will become conscious. Even if this proves to be nonsense, we may find that our collective Shen is sufficient to protect Earth from the Sun's cosmic radiation just as the meditators in the experiment described in Chapter 4 were able to prevent cosmic radiation from passing through a shielded room.

Just as extensive prayer or meditation can change a person's Kirlian image from Qi (left) to Shen (right) perhaps extensive prayer and meditation can change Earth's electromagnetic Qi field into a Shen field.

Is it possible that we, through our own efforts, are taking all the "negative" electrons of the world, and transmuting them to Shen, a much more positive, spiritual, and conscious substance? Perhaps Qigong, and all forms of prayer and meditation are actually the cause of our transition into a New Age, dominated by Shen, a higher consciousness. *Just maybe, our thoughts do make a difference.*

Qigong stands as a bridge, not only as a pathway for individual growth, but also as an opportunity for unprecedented scientific discovery. Kirlian photography provides a critical bridge for western scientific understanding of Qigong science. The new field of Bioelectric Vitality empowers you to make a difference.

Key Points from Chapter 6:

1. We have weakened Earth's Qi field with our technology, depleting the ozone layer with airborne pollutants. We can help to strengthen Earth's Qi field by mandating the attachment of powerful ionizers to airplane wings and cars. This will both increase ozone in the ionosphere and go far to reducing air pollution in our cities.

2. Qigong breathing may be able to increase the negative charge of Earth, fortifying the ozone layer by drawing electrons from the heavens. It would probably serve humanity well to subject the mechanism of Qigong breathing to extensive scientific study.

3. Shen Gong, and other prayer and meditation methods are practiced extensively around the world. Shen Gong exercises may consume free electrons, *decreasing* the electrical charge of Earth as Shen is produced. These practices may also be creating a very strong worldwide Shen field which may change the nature of life on Earth. Again, scientific study is in order!

BIBLIOGRAPHY

Becker, Robert O. *Cross Currents: The Perils of Electropollution*, Los Angeles: Jeremy P. Tarcher, Inc., 1990. Lays the scientific groundwork for energy medicine, providing insight into the medical benefits of electrical currents and magnetic fields in healing, and the dangers to our health of electromagnetic fields produced by our technology.

Dumitrescu, Ion, and Julian Kenyon. *Electrographic Imaging in Medicine and Biology*, Suffolk, England: Neville Spearman Limited, 1983. Presents research that streamers appear around hands of cancer patients and balls form around the cancer site itself. This book is a wealth of information on Kirlian photography and many other forms of high voltage electrical imaging.

Hunt, Valerie. *Infinite Mind: The Science of Human Vibration*, Malibu: Malibu Publishing Company, 1995. Presents a lifetime of important research, including discussion of chaos electromagnetic radiation measurable at the chakra points on the human body which correlate with clairvoyant view of the body.

Konikiewicz, Leonard, and Leonard C. Griff. *Bioelectrography, A New Method for Detecting Cancer and Monitoring Body Physiology*, Harrisburg: Leonard Associates Press, 1984. Presents many years of clinical research on Kirlian photography providing insight into the Kirlian phenomenon.

Maciocia, Giovanni. *The Foundations of Chinese Medicine*, New York: Churchill Livingstone, 1989. An easy-to-understand textbook on traditional Chinese medicine which provides deeper understanding into many aspects of the human energy body.

Mandel, Peter. *Energy Emission Analysis*, W. Germany: Synthesis Publishing Company, Siegmar Gerken. A compilation of a vast amount of clinical research into the significance of Kirlian photographs in terms of meridians and acupuncture theory. Presents an interesting theory of clinical diagnosis using Kirlian photographs of fingers and toes.

Soyka, Fred with Alan Edmonds. *The Ion Effect, How Air Electricity Rules Your Life and Health*, New York: Elsevier Dutton Publishing, 1981. A wealth of information on the biological effects of negative ion deficiency, the negative effects of weather patterns involving descending air masses, and the benefits of ion generators.

Talbot, Michael. *The Holographic Universe*, New York: Harper Collins, 1991. This book provides a new framework for understanding the phenomenon of Qigong and all other paranormal phenomenon in terms of the hologram. A very valuable work.

White, John. *Pole Shift*, Virginia Beach: ARE Press, 1986. A classic prediction of impending doom, most of which didn't happen, but which still might. A fascinating collection of geophysical mechanisms waiting to spring into activity.

Index

About the Author

Richard H. Lee a native Californian, received his Bachelor of Science degree in Systems Engineering from Harvey Mudd College, and is a Registered Mechanical Engineer (P.E.). For over a decade Mr. Lee performed energy conservation surveys and engineering design modifications for heating and air conditioning systems, saving his clients millions of dollars in fuel costs. He is a noted author in the field, has presented papers at international energy conferences, and holds U.S. and international patents.

In 1989 Mr. Lee turned his sights to human energy when he founded China Healthways Institute. Like any typical western scientist, he questioned how these traditional Chinese therapies worked. Were Chinese herbs just vitamins? Were successful results with acupuncture simply due to the placebo effect? And what about Qi, the Chinese concept of vital energy which was the central tenet of Chinese medicine? Was Qi real, or just a metaphor?

Mr. Lee has long since answered his own questions, but now spends his time in research and study in order to validate Qi based therapies scientifically for the western world. Medical infrasonics and bioelectric analysis have been his principal research tools. China Healthways Institute has been his vehicle.

China Healthways Institute is dedicated to educating western health care providers about the importance of vital energy (Qi) in health and patient recovery, believing that incorporating an understanding of the energetic nature of healing will greatly enhance the western health care system, reducing both side effects and medical costs.

China Healthways is inaugurating a book publishing program with BIOELECTRIC VITALITY to complement its popular publication of a quarterly newsletter. CHI books focus on in-depth explorations of issues of interest to eastern and western health care professionals.

e-mail: chi@exo.com

Home page: http://exo.com/~chi/

phone: (714) 361-3976